Hop On the Water Cycle

水无处不在 水的循环

[美] 纳迪亚·希金斯 著

[美] 莎拉·英芬特 绘

[美] 德鲁·登贝朗德 作曲

池欣阳 译

中国水利水电出版社
www.waterpub.com.cn
·北京·

项目策划：徐丽娟
责任编辑：栾 峰 方 斯
特约编辑：李渝汶
联系方式：luanfeng@mwr.gov.cn 010-68545978

书　　名	水无处不在　水的循环 SHUI WUCHUBUZAI　SHUI DE XUNHUAN
作　　者	［美］纳迪亚·希金斯 著　［美］莎拉·英芬特 绘 ［美］德鲁·登贝朗德 作曲　池欣阳 译
出版发行	中国水利水电出版社 （北京市海淀区玉渊潭南路1号D座　100038） 网　　址：www.waterpub.com.cn E-mail：sales@mwr.gov.cn 电　　话：（010）68367658（营销中心）
经　　售	北京科水图书销售中心（零售） 电　　话：（010）88383994、63202643、68545874 全国各地新华书店和相关出版物销售网点
排　　版	陆　云
印　　刷	北京尚唐印刷包装有限公司
规　　格	285mm×210mm　16开本　6印张（总）　80千字（总）
版　　次	2022年1月第1版　2022年1月第1次印刷
总 定 价	148.00元（全4册）

图书在版编目（CIP）数据

水无处不在．水的循环：汉英对照／（美）纳迪亚
·希金斯著；池欣阳译．-- 北京：中国水利水电出版
社，2022.1
书名原文：Water All Around Us
ISBN 978-7-5226-0105-2

Ⅰ．①水… Ⅱ．①纳… ②池… Ⅲ．①水—儿童读物
—汉、英 Ⅳ．① P33-49

中国版本图书馆 CIP 数据核字（2021）第 210677 号

北京市版权局著作权合同登记号：图字 01-2021-5364

亲子学习诀窍

为什么和孩子一起阅读、唱歌这么重要？

每天和孩子一起阅读，可以让孩子的学习更有成效。音乐和歌谣，有着变化丰富的韵律，对孩子来说充满乐趣，也对孩子生活认知和语言学习大有助益。音乐可以非常好地把乐感和阅读能力锻炼有机结合，唱歌可以帮助孩子积累词汇和提高语言能力。而且，在阅读的同时欣赏音乐也是增进亲子感情的好方式。

记住：要每天一起阅读、唱歌哦！

绘本使用指导

1. 唱和读的同时找出每页中的同韵单词，再想想有没有其他同韵单词。
2. 记住简单的押韵词，并且唱出来。这可以培养孩子的综合技能以及英语阅读能力。
3. 最后一页的"读书活动指导"可以帮助家长更好地为孩子讲故事。想一想，音符和歌词里的单词有什么联系？
4. 跟孩子一起听歌的时候可以把歌词读给孩子听。
5. 在路上，在家中，随时都可以唱一唱。扫描每本书的二维码可以听到音乐哦。

每天陪孩子读书，是给孩子最好的陪伴。

祝你们读得快乐，唱得开心！

扫我听音乐

Water keeps going around and around! This is called the water cycle. The water cycle is made up of three steps. Evaporation is when water vapor rises into the air. Condensation is when water droplets gather into clouds. Precipitation happens when water falls from the clouds.

Turn the page to learn more about the water cycle.
Remember to sing along!

水总在旅行，叫作水循环。
构成水循环，环节有三段。
第一是蒸发，水汽升空中。
第二是凝结，水珠聚成云。
第三是降水，雨雪云中落。

现在，请翻到下一页，
我们一起认识水的循环吧！跟着音乐一起唱吧！

Hop on the water cycle.
It's the wettest ride in town:
water changing, water moving
up and down and all around.

水的循环趣味多，整个旅行湿漉漉，
水在变身，水在移动，
天地之间，四处弥漫。

Grab your umbrella. Step right up!

带上雨伞，这就出发！

Ready, set, first stop: evaporation!

做好准备！第一站：蒸发！

Shining down, the sun's hot rays turn water into its gas phase.

烈日炎炎当空照， 水变气态静悄悄。

From oceans, puddles, anything wet,
water vapor floats over your head.

海洋，水坑，潮湿处， 水汽空中飘呀飘。

Hop on the water cycle.

It's the wettest ride in town:

water changing, water moving

up and down and all around.

水的循环趣味多，整个旅行湿漉漉，
水在变身，水在移动，天地之间，四处弥漫。

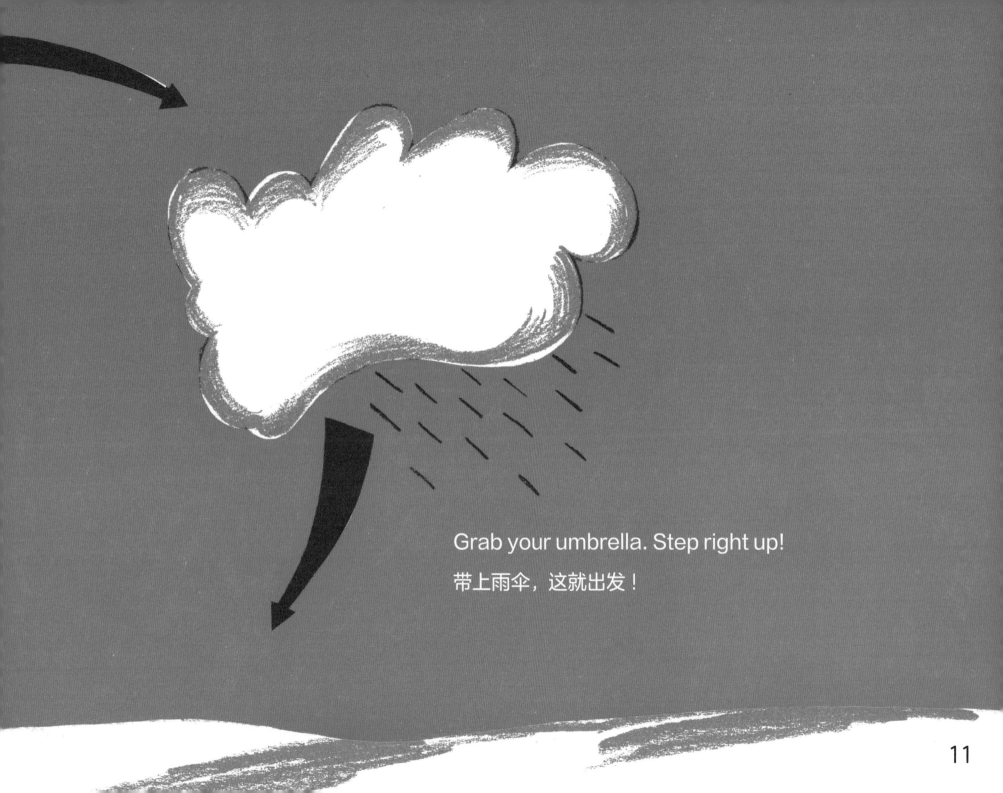

Grab your umbrella. Step right up!

带上雨伞，这就出发！

Ready, set, second stop: condensation!

做好准备！第二站：凝结！

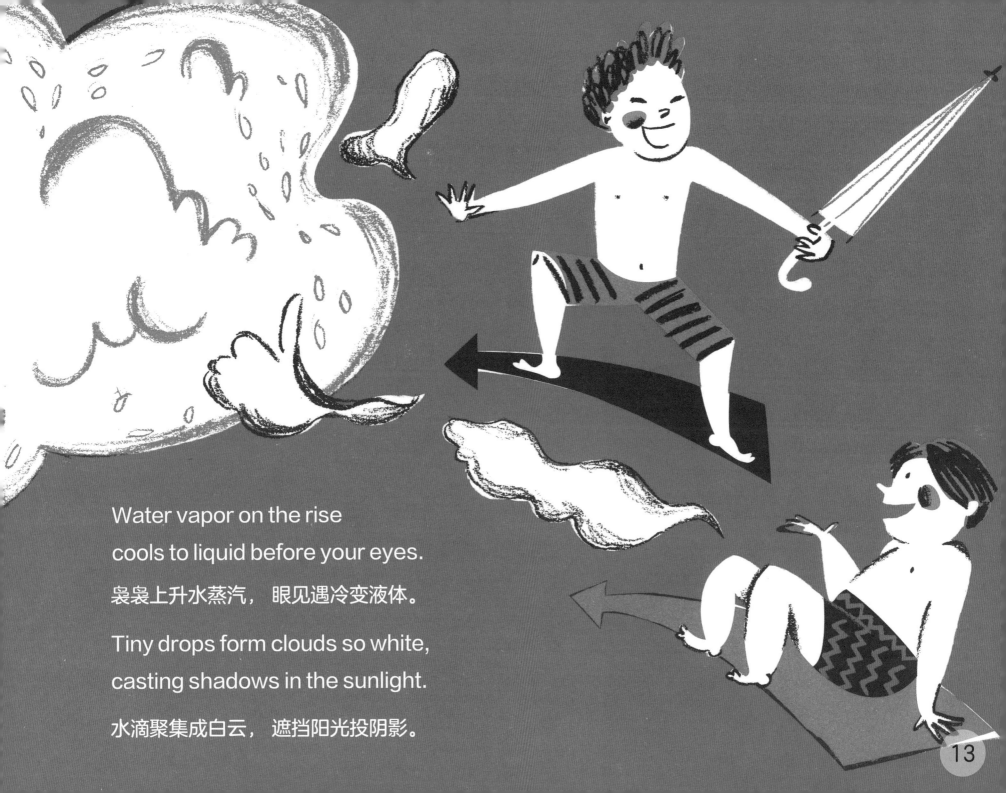

Water vapor on the rise
cools to liquid before your eyes.

袅袅上升水蒸汽，　眼见遇冷变液体。

Tiny drops form clouds so white,
casting shadows in the sunlight.

水滴聚集成白云，　遮挡阳光投阴影。

13

Hop on the water cycle.
It's the wettest ride in town:
water changing, water moving
up and down and all around.

水的循环趣味多，
整个旅行湿漉漉，
水在变身，水在移动，
天地之间，四处弥漫。

Grab your umbrella. Step right up!

带上雨伞，这就出发！

Ready, set, third stop: precipitation!

做好准备！第三站：降水！

In a cloud, drops bump and grow,
oh, so heavy. They cannot float.
水滴碰撞汇一体，哎呀，好重！飘不起。

Down comes rain and down comes snow.
Back to oceans water flows.
降落成雨，飘飞成雪。回归海洋波涛里。

Dance stop! Do the water cycle bop!

舞蹈站到啦！跟着音乐跳水循环之舞吧！

Evaporation, float your hands in the air.

Condensation, shape clouds in your hair.

Precipitation, bring rain to your feet.

Keep moving with the water cycle beat!

大家一起来蒸发，小手在空中挥舞。
大家一起来凝结，用头发编织云朵。
大家一起来降水，雨点掉在脚背上。
跟着水循环的节奏，大家动起来！

Hop on the water cycle.
It's the wettest ride in town:
water changing, water moving
up and down and all around.

水的循环趣味多， 整个旅行湿漉漉，
水在变身，水在移动， 天地之间，四处弥漫。

Hop on the water cycle.
It's the wettest ride in town:
water changing, water moving
up and down and all around.

水的循环趣味多，整个旅行湿漉漉，
水在变身，水在移动，天地之间，四处弥漫。

SONG LYRICS 歌词
Hop On the Water Cycle

Hop on the water cycle.
It's the wettest ride in town:
water changing, water moving
up and down and all around.
Grab your umbrella. Step right up!

Ready, set, first stop: evaporation!
Shining down, the sun's hot rays
turn water into its gas phase.
From oceans, puddles, anything wet,
water vapor floats over your head.

Hop on the water cycle.
It's the wettest ride in town:
water changing, water moving
up and down and all around.
Grab your umbrella. Step right up!

Ready, set, second stop: condensation!
Water vapor on the rise
cools to liquid before your eyes.
Tiny drops form clouds so white,
casting shadows in the sunlight.

Hop on the water cycle.
It's the wettest ride in town:
water changing, water moving
up and down and all around.
Grab your umbrella. Step right up!

Ready, set, third stop: precipitation!
In a cloud, drops bump and grow,
oh, so, heavy. They cannot float.
Down comes rain and down comes snow.
Back to oceans water flows.

Dance stop! Do the water cycle bop!
Evaporation, float your hands in the air.
Condensation, shape clouds in your hair.
Precipitation, bring rain to your feet.
Keep moving with the water cycle beat!

Hop on the water cycle.
It's the wettest ride in town:
water changing, water moving
up and down and all around.

Hop on the water cycle.
It's the wettest ride in town:
water changing, water moving
up and down and all around.

Hop On the Water Cycle

Verse 2
Ready, set, second stop: condensation!
Water vapor on the rise
cools to liquid before your eyes.
Tiny drops form clouds so white,
casting shadows in the sunlight.

Verse 3
Ready, set, third stop: precipitation!
In a cloud, drops bump and grow,
oh, so, heavy. They cannot float.
Down comes rain and down comes snow.
Back to oceans water flows.

GLOSSARY   词汇表

condensation—when tiny water droplets gather to form clouds

凝结——细小的水珠汇聚在一起形成云朵

cycle—a set of things that happen again and again in the same order

循环——事情按照固定的顺序重复发生

evaporation—when water heats up and turns from liquid to water vapor

蒸发——液态的水受热后变成水蒸气

gas—something that is like air and has no shape

气态——像空气一样没有固定形状的物质形态

precipitation—when water falls from clouds as snow, rain, hail, or sleet

降水——雨、雪、冰雹或雨夹雪从云中落下的过程

读书活动指导

1. 想像一下水的循环。我们能看到循环的哪些部分？又有哪些部分我们看不到？

2. 水的循环由蒸发、凝结和降水三个环节组成。选择其中一个环节画一画吧。

3. 水的循环有着一系列的步骤。想想我们生活中每天都要做的事情，比如上学、放学、午餐等。选择其中的一件，列一列需要的步骤吧。